Christine Roux is the President of I.T. By The Sea, LLC. She has been working in the IT field for over twenty-five years helping home users, small and medium businesses with their computers and their networks.

She was born and raised in Paris, France and holds three Masters' Degrees in French Law, Criminal Sociology, and Information and Communication Sciences from the Pantheon Sorbonne University in Paris. She passed with Honors the Paris Bar, and worked as an Attorney in Paris until she decided to relocate to Buenos Aires, Argentina.

In the early 1980s, she was teaching English as a Second Language for a company in Buenos Aires, Argentina whilst she was studying to revalidate her degrees. One day, her classroom computer broke ten minutes before she had to begin teaching a class consisting of eight high level executives. She looked around and realized it was up to her to fix the computer. Not only did she manage to do it, with the help of a couple of friends on the phone, but she also discovered that she was fascinated with computers. A month later, she was fixing all the other teachers' computers. Six months later, she was taking care of the whole corporate network and she kept administrating and managing it for over fifteen years.

In 1997, she came to the U.S. and went back to school, at age 45, to learn more about the "whys" and the "hows" of computers and networks.

Christine now holds a Bachelors' degree with a major in Computer Information Technology (CIT), and a Masters' degree in CIT with a Specialization in Networking. She did all her graduate work, including her Masters' Project on Networking and Virtualization.

She also holds several Vendor and Vendor-neutral certifications: COMPTIA A+, COMPTIA Network+, COMPTIA Security+, COMPTIA IT Project+, Microsoft Certified Professional, Microsoft Certified Database Administrator, Microsoft Certified Systems Administrator, and Microsoft Certified Systems Engineer.

Christine is perfectly trilingual, with English being her third language after French and Spanish. She also studied Italian, German, Portuguese, Arabic, Quechua and Guarani. She is a gifted public speaker who manages to both educate and entertain even when dealing with the driest technical topics.

Christine started I.T. By The Sea, after several years in the academic world, because she was appalled by the stories she was hearing about the way so many computer technicians treat their clients. She wanted to create a company that made it clear to all their clients that they cared about them, about their computers, about their networks and about their business.

She is proud to say, "That is what our clients actually say about us."

i

ii

A Hurricane

It's A Coming

How to Prepare and Execute Disaster Recovery,
Business Continuity and Emergency Survival Plans
To Ensure Your Business Stays in Business After a Storm

By

By Christine F. Roux

MCSE, MS CIT

www.ITbytheSea.com

Roux, Christine

  A Hurricane It's a Coming: how to prepare and execute disaster recovery, business continuity and emergency survival plans to ensure your business stays in business after a storm

ISBN 1453825266

Publisher: I.T. By The Sea, LLC
             13102 Palm Beach Boulevard
             Suite A
             Fort Myers, FL 33905

Illustrations by J. Walgren

DEDICATION

To Albert Ball

Program Chair of Computer Information Technology at Hodges University

Thank you for all the knowledge you so selflessly shared with me.

FOREWORD

I first met Christine Roux in January 2001 when I was the Program Chair of Computer Information Technology at International College, now Hodges University, in Naples/Fort Myers, Florida. At the time, she was entering the college as both a student and a Teaching Assistant.

Since then I have had numerous opportunities to deal with her and to evaluate her work as a student, a Teaching Assistant, a College employee, then an Instructor and a contractor. I have always found her work to be extremely thorough, logical and well organized. I also know that she has always enjoyed all the internet research involved in both her studies and her position as an Instructor. This is why I was not surprised to see that her book, *A Hurricane It's A Coming,* was based on deep internet and offline research on hurricane preparedness for businesses.

I had to laugh when I read that Christine not only recommended to bag electronic equipment to protect it against a storm but even offered a video on how to do it. As one of the employees in charge of our classrooms technology, Christine demonstrated a wide experience in "computer bagging." Remember that Hodges University is located in South West Florida, a region that has been under numerous hurricane alerts in the past few years.

I especially remember one year when it felt like we were having one storm alert after another. Christine, of course, participated in the repeated taking apart and setting up of the computers and network devices for each storm alert. At the end of the storm season, she was asking if there was a Guinness Record for "Computer Bagging" because she thought the school team would win it easily thanks to the amount of practice nature had just given them!

A Hurricane It's A Coming is very easy to read and even easier to act on. As the College Instructor she used to be, Christine teaches you, her fellow business owners, step by step what you need to do to protect your business against the possible consequences of a hurricane. She even provides check lists that will help you ensure that all the angles have been covered.

If you are a business owner in hurricane country, you not only need to read this book, you need to act on it. Use the notes sections to write down your own ideas; use the checklists to make sure you are as prepared as can be; write up your own Business Continuity and Disaster Prevention Plans. And, don't forget to bag your equipment although hopefully nature will never give you enough practice to even think of competing for a Guinness Record in "Computer Bagging!"

David Rice
Vice President of Information Technology & Facilities Management
Hodges University – Naples/Fort Myers, Florida.

Table of Contents

INTRODUCTION

If you have been living anywhere around the Gulf Coast or on the East Coast of Florida for more than a couple of days between June and November there is one thing you know for certain... It is Hurricane Season and **THE HURRICANES ARE COMING**.

You are a successful entrepreneur; you've put your heart and soul into your business. You have given it your 110%. You have sacrificed your time, spent your energy and invested your money to build your business and make your dreams come true. Now, a storm is coming that could destroy your business and all your hard work. You board up the windows, sandbag the doors and secure things as best you can. You take cover and ride out the storm. After the storm you breathe a sigh of relief, All is Well... OR IS IT?

The purpose of this book, *A Hurricane It's a Coming*, is to teach you Step-by-Step how to prepare and execute Disaster Recovery, Business Continuity and Emergency Survival Plans to ensure your business stays in business after the storm.

Why You Need This Book

Research indicates that twenty percent of businesses that close due to the effects of hurricanes and other natural or manmade disasters never reopen their doors even when their physical facilities survived, even when they thought they were prepared.

WHY? *A Hurricane It's a Coming* explains why.

As a business owner, you may have done research to find instructions and advice on how to best protect your business so that it has a chance to survive a storm even if you take a direct hit. That's probably why you're looking at this book right now. And, you most likely found exactly what we found: lots of books and internet sites with an overload of information but not one single place on which you can rely to speak to all the topics hurricane

Write Notes Here:

preparedness for business involves. This is the reason WHY I wrote this book.

Do I have services to offer you during this season? I most certainly do. My Company, I.T. by the Sea (http://itbythesea.com), is an Information Technology company, and we will be glad to help you get ready for the season or help you get your business back on its feet should you get hit.

We can help you document your Network, prepare your Disaster Recovery and Business Continuity Plans, and get ready for the storms. We also offer home and business assets inventory through our Florida Inventory Services http://FloridaInventoryServices.com division and can help you make sure that you are appropriately insured and have all the proofs your insurance company will ask for if disaster strikes. Finally, if you are unlucky enough to get hit by a storm, we can help you recover and get your technology back up in the shortest possible time. Our help is never further than an email (info@itbythesea.com) away.

However, I did not write this book for the sake of offering my services. My main objective was to share the information I have and help you get as prepared as possible. If I can bring some peace to your mind during a hurricane season, I will be happy.

Now, I would like to make it clear that I am not a special expert on the topic of hurricanes. However, I do tend to know a lot more than most people thanks to our Florida Inventory Services division that has kept me in constant contact with insurance and disaster recovery companies. I also know more than most people about the best way to protect your technological devices because technology has been my daily companion and the focus of my education for the past twenty-five years.

My credentials include: a Bachelor's Degree and a Master's Degree in Computer Information Technology and several certifications: A+, Network+, Project+, Security+, MCP, MCSA, MCSE and MCDBA. I am a member of The ASCII Group and The ACRBO (Association of Computer Repair Business Owners). I am an expert in researching, especially on the internet. My research on the topics addressed in this book has been long and extensive. I have done my best to bring together all the knowledge I have garnered and present it to you all in one place, in a concise, useable manner.

Can I promise you that this book will keep your business safe even from a direct hit? I wish I could, but as we all know, nobody could possibly make and keep such a promise. The only thing I can promise you is that, if you read this book, watch the videos, and do everything experts advise us to do, you will have given your business its best chances to survive even a direct hit.

What's In It For You?

Register on this book's website: (http://ProtectYourBusinessFromHurricanes.com) to access bonus videos, resources and downloadable copies of the worksheets and checklists found in the appendix at the back of this book. It is my hope that you will use these bonus resources to help you prepare and update the information you will need to develop your Emergency Survival, Disaster Recovery and Business Continuity Plans.

Serious Silly Talk

It is human nature to delay preparation for catastrophic events over which one feels they have no control. Instead of preparing, apathy kicks into high gear; people become overwhelmed and stop thinking logically. Then, they start to do crazy things. Research indicates that eighty percent of businesses worldwide are not prepared to handle or survive natural disasters and other catastrophic events (even when they are aware of the possibilities and preparation could make a significant difference).

In *A Hurricane It's a Coming*, I have included some **Serious Silly Talk** to inspire you to act, and for you to use to educate, enlighten, encourage and inspire your employees, customers, vendors, suppliers and all other people your business depends on to take action to prepare for emergencies. You need to get them to take action because your business survival depends, to a great extent, on their survival. Any weak link in the chain could spell disaster. Survival success requires the cooperation of every link in your business chain; it requires ACTION!

Keeping it light and adding a laugh by studying a real or an imaginary third party will reduce fear and help deter paralysis when action is required.

How This Book is Organized:

A Hurricane It's a Coming is written like a textbook with large margins on the outside of each page for you to use to jot down your thoughts and take notes as you read. The book is organized in three parts:

Part One - Before the storm
Part Two - When a storm is coming
Part Three - After the storm

Each part is broken into three sections:

Section 1 - gives general information
Section 2 - gives business specific information
Section 3 - gives technology specific information

There is a TAKE ACTION WORKSHEET at the end of each section that you can use to organize your section notes. Later, you will use the Take Action Worksheets to help you develop your personalized Disaster Recovery, Business Continuity and Emergency Survival Plans. I have also provided links to online EMERGENCY CHECKLISTS that you can download when you are ready to develop your plans. After you have completed filling in your Emergency Checklists, you can print them and use them in your desk manual during emergencies. You can also save a copy of them on a CD or jump drive to put in your GO BAG.

Procrastination Is Suicide On The Installment Plan

~Robert Allen

NOTES

Part 1

BEFORE THE STORM –

Preparing for Emergencies

Serious Silly Talk ~

Who Can Sell The Most Hamburgers?

A group of Florida restaurant owners was enjoying a friendly round of golf when one started bragging about the massive volume of his hamburger sales. As he was talking, he pulled a large wad of cash out of his pocket to flash before his friends. This made the others in the group very jealous. As nature would have it, another owner countered by bragging about the high quality of food his restaurant served. And, another owner bragged about the variety and creativity of his menu. Before long everyone in the group had joined the boastful competition declaring their own restaurant to be the best and most profitable business because of one attribute or another.

"Are you willing to put your money where your mouth is?" one friend asked the restaurant owner with the large wad of cash.

"I've got five grand in my pocket!" the owner answered as he flashed the cash again for all his friends to envy. "I'll bet this whole wad of cash that I can sell more hamburgers and make more profit than any of you. And, when you loose, you'll each have to pay me five grand... deal?"

Everyone said, "Deal!" and agreed on a month long contest that would begin the first day of August.

The first owner had a prime location with lots of drive-by traffic. He was sure he could win the $20,000 prize. All he needed to do was advertise a special on his large sign out front. Four Hamburgers for $5 would do the trick.

The second owner was sure that the quality of his Black Angus hamburgers would produce the needed sales to win the contest. He decided to advertise using a tempting photograph of his deliciously beautiful hamburger piled high with red leaf lettuce, pickles and onion on a toasted sesame seed bun. Just thinking about it made his mouth water.

The third owner was proud of the variety on his menu. He decided to feature a different one of his Deluxe Burgers each week starting with the "Super Deluxe Mile-High Bacon and Cheddar Mushroom Burger smothered in a special secret sauce."

The fourth owner was proud of the relationship he had developed with his employees. He was sure that his loyal staff would be the deciding factor…

Every owner had a unique strategy that they believed would help them win the prize. Each owner had sacrificed a great deal of time, energy, money and creativity to develop their business. Their businesses had become their lives and provided the means to sustain their families while they lived out their dreams. They could play golf any time they wanted to. They owned yachts and planes and could travel anywhere in the world. Not only that, the businesses they owned provided the means to sustain the families of their many employees. Lots of people were depending on them and they were very proud of that fact.

Of course each owner thought that they would win. Why else would any of them risk so much of their hard-earned money?

Question: What do you think it would take to win the competition?
- o Good Food
- o The Best Ingredients
- o Variety and Creativity in the Menu
- o High Traffic Volume
- o Great Service
- o None of the Above

Congratulations if you answered, "None of the above."

What you need to win under NORMAL CIRCUMSTANCES is:
correct pricing and the greatest number of very hungry customers.

_____
_____
_____
_____
_____
_____
_____
_____
_____
_____
_____
_____
_____
_____
_____
_____
_____
_____
_____
_____
_____
_____
_____
_____
_____
_____

However, the name of this book is _A Hurricane It's A Coming,_ and come she did. All of the restaurant owners were equally affected by the storm. They boarded up their windows and secured their doors. The owners and their employees evacuated the area taking their families with them. High winds downed trees and power lines. Heavy rains caused flooding; however, none of the restaurants sustained any serious structural damage.

After the storm the restaurant owners returned to survey the damage. All of them faced the same challenges to reopen their businesses so they agreed to continue their competition.

Question: What do you think it would take for the restaurant owners to win the competition after the storm?

- _____
- _____
- _____
- _____
- _____
- _____
- _____
- _____
- _____
- _____
- _____
- _____
- _____
- _____
- _____
- _____
- _____
- _____

There are dozens of answers and ALL of them are probably correct to some degree or another. The problem lies not in the answers with which you may have come up. Trouble comes in the answers you didn't think of. If there was any answer, any detail that was NOT considered, addressed and prepared for in advance, the contest and the owners' restaurants could be lost along with the livelihood of the restaurant owner and all of his employees. It is important to brainstorm until you think of every necessary detail.

Here is A Sobering Fact to Consider:

Statistics reveal that 20% of the businesses that close after a catastrophic event never reopen their doors.

Part 1: Before the Storm

The first point I would like to make is that before the storm does not mean the day the TV tells you there is a storm coming your way. Before the storm means NOW!!!

Please, make sure you do not wait until the stress is hitting the fan. If you do, you might not be able to obtain critical supplies, or you might miss or forget some important precautions and thus endanger your life, the lives of your employees and the survival of your business.

Questions:

When should you start to prepare? _____

What are the consequences of NOT being prepared? _____
_____
_____
_____
_____
_____

What are the advantages of preparing NOW? _____
_____
_____
_____

Are you committed to preparing now? _____
_____
_____
_____

Why or why not? _____
_____
_____
_____
_____

Write a mission statement for your Disaster Recovery, Business Continuity and Emergency Preparedness Plans: _____
_____
_____
_____
_____
_____
_____
_____
_____
_____
_____
_____
_____
_____
_____
_____
_____
_____
_____
_____
_____
_____
_____
_____
_____
_____
_____

Section 1: General Information

The first thing you need to do is to try and prevent the storm from penetrating your facility. We all know there is not much we can do to prevent trees from falling on the roof although cutting any and all prominent branches, or contracting someone to cut them would not be a bad idea. You might also want to have your roof inspected to make sure there is not a small leak that a big storm could transform into a cataract. Covering your windows and doors, on the other hand, is a must. Of course, you will not cover them right now, but you need to get prepared immediately so that you do not have to run to the store at the last moment… when everyone else is there too.

Openings and Coverings

Even if the threat is a tropical storm and not a hurricane, there is the hazard of high and dangerous winds. The definition of a tropical storm is: sustained winds of 39 to 73 MPH. With that wind speed, a coconut can become a missile. This is why you should protect the facility's openings when preparing for a hurricane or tropical storm.

Doors

When you prepare to protect your facility's openings, do not forget to take into account all the external doors of the facility. Even if the doors of your business do not include a glass part, which they often do, you need to make sure they will be strong enough to live through the storm. You do not want to take all the other precautions only to discover after the storm that the doors let you down and your office or business has been destroyed.

List the numbers of external doors you need to protect in the margin. Go ahead. Do it NOW!

Windows

Your first step in protecting your windows will be deciding how you want to protect them. If you rent your office space, you will most likely have to provide your own window and door protections. Do you want to board up your windows, or do you want to use roll down shutters or corrugated metal to protect them?

If you chose boarding up your windows, Plywood is the most economical, but if storage is an issue you may want to consider corrugated metal. It is lighter in weight than plywood and comes in stackable smaller sheets.

If you are choosing to go with plywood, note that the plywood needs to be 5/8"-3/4" exterior plywood. Cut each shutter according to the size of the window. To hang the shutters, you will need either clips made for this purpose or 3"- 4" heavy-duty barrel bolts. Do not forget to purchase them and place them in a safe location. Again, you don't want to have to run to a VERY busy store at the last moment. You may also want to paint these shutters to protect the plywood from the weather.

Whichever way you decide to protect your windows and glass doors, planning is the key. The time to do it is now! During a threat of a hurricane is NOT the time to purchase the proper materials or contact your local storm shutter company. The plywood and corrugated sheeting is difficult to find and all the procrastinators will be running around like crazy to find enough material to protect their windows.

After purchasing your window protection, be prepared to spend a weekend measuring and cutting your material. You should also mark the covering for each opening, so when the time comes to implement your glass defense, you will not have to rely on your memory as to where things go. If you already own protective opening coverings, you need to check to make sure everything is in good condition. Do a rehearsal on putting up your plywood or shutters before the hurricane. This can really help relieve any additional stress should you receive a hurricane threat.

Alternative Location

Should there be structural damage to the facility where your business is presently located, you will need to find an alternative location. Again, do not leave the hunt and the choice for the last minute, or even worse for after the storm. The time to visit alternative locations, choose one or two and discuss conditions with your potential landlords is NOW.

You need to consider a place that is not too close to your current facility because if you get hit so will a place around the corner. You don't want it to be too far away either because you need to be able to keep your clientele. You will need to inspect the locations outside and inside. Make sure there will be enough space for all employees.

You also need to make sure the network cabling is healthy and offers enough drops to accommodate all your computers. Make sure the electrical installation is strong enough to easily align all your electronic devices. Choose a facility as comfortable and practical as possible. (See the Business Specific section of this chapter for additional advice on how to choose the alternative location so that the temporary move hurts your business as little as possible.)

Ideally, the alternative location would be a "hot site" i.e. a site where you could just move in and keep doing business as usual because everything you need to keep your business running would already be in place. Unfortunately, we all know that this is completely out of the economic reach of most small and medium businesses. So, unless you're one of the happy few, if your business receives structural damage during a hurricane you will have to find another place to set up shop. When choosing another location it is important to keep your businesses needs in mind.

One good way to get an alternative location is to work with business to business planning. Make a deal with another business that if you get hit, your company works out of their place and if they get hit, their company will work out of your place. Of course, you will have to seek a business that is not only non-competitive and friendly but also located at a certain distance from your current facility to decrease the chances of both businesses being hit by the same storm.

Another way to identify possible alternative locations is to take a ride one afternoon and scout out some places that are for rent. This is one thing you do not want to do too many months in advance because the place could be rented or change owners by the time you might need it. Jot down the pertinent information such as where it is located; the phone number of the contact person, whether that be a realtor or a private owner; and the contact person's name. It is also a good idea to take a picture of the place. Give the owner a phone call, introduce yourself and your business and let them know what you are doing. Be sure to put this information in the information binder you have in your firebox. Put this list in order of your first choice, second choice, etc. You need to be sure your staff members are aware of these locations and any special instructions that go along with these locations.

You will need the full address of the alternative location. Be sure all your employees are aware of the alternative business location and know what to expect should you have to use it in the future. If you do need to use the alternative business location in the future, be prepared to work with minimal means. The new space may have less room, office sharing, or no internet for a while (unless it was not affected by the storm). These are just a few of the examples.

Utilities

Make sure you know where the water, gas and power main cutoffs are located in your own facility, and in the alternative locations you choose, and document it! This is something you should be aware of even if you do not live in hurricane country. You should also communicate this to your employees. You should also know how to turn these utilities off and back on, and you should share this information with your employees as well.

Phones

Your phone landlines may be down for a while. This does not mean your cell phone will not work or vice versa. Know how to forward your phone because you may have to use your cell phone as your business line for a few days. In fact, if your office has structural damage, you may have to forward your business line to your cell phone until the proper repairs are made to your building. If you have to use an alternative facility, it might also be easier and cheaper to use your cell phone as your business line than to have a new line connected in the alternative location in the middle of the after storm chaos. You will also want to have a NON-cordless phone. A cordless phone will not charge or work if there is not electricity. Be sure to have extra cell phone batteries just in case you need to use your cell phone as your business line.

Generator and Fuel

If your business is hit by a hurricane you need to be prepared to work with minimal means. This could mean without electricity or running water. Having a backup generator would be very helpful if your office or shop is not damaged but the electrical lines are down. If you already have a generator, you will want to make sure it is in good running condition. If you have used the generator in the past, and stored it with gas in it, the carburetors have a tendency to get gummed up. This can happen even if you have put an additive in the gas when you stored it. Be sure to contact a licensed generator mechanic to have your generator checked before the threat of a storm.

You will also need extra storage for the fuel for your generator. These containers need to be specifically approved containers for the exclusive use of fuel storage. An old plastic milk jug is NOT an approved gas container. Gasoline containers are readily available in hardware stores. Be sure to get these before the threat of a hurricane because once there is a named storm, these containers are scooped up quickly and become quite scarce.

Miscellaneous

Have a spare **tarp** or two. If part of your roof is damaged, you will be happy to be able to cover it by nailing down a tarp on the destructed area to stop the rain from getting into your building, even if you are using an alternative location. Of course, to do so, you will also need a hammer, nails and a ladder.

Tip: Better than tarp alternative - Some advertising sign companies sell their used billboard signs for a very reasonable price. (around $25.00). The signs are made of an extremely durable heavyweight waterproof vinyl and usually measure 15'x 50'.

It is also a good idea to make sure you have abundant supplies of flashlights, lots of trash bags and extra batteries for all the battery-operated items you own. Whilst you are purchasing all those supplies, remember to add a couple of fix-a-flats to your basket. After a storm there will be lots of nails, glass parts and debris on the ground and they can be lethal to any car or even truck tires.

First Aid Kit

There may be nails, broken glass, or a hundred other ways to get hurt after a hurricane. Your business should already have a stocked first aid kit but now would be a good time to check it out. Sometimes items get old and dried out. Take the time to replace missing or old medical items.

Shoes and Gloves

Common sense dictates that each employee should own, and have available, a good pair of shoes with hard soles and a pair of work gloves to wear when preparing for and cleaning up after a hurricane.

Before the Storm General Information Brainstorming and Note Summary

Section 2: Business Specific Information

Business Continuity and Disaster Recovery Plans

As you most certainly know, you should already have both a Business Continuity Plan and a Disaster Recovery Plan in place for your business. The key to successful disaster recovery is to have them in place well before disaster ever strikes. Business advisors have been giving us this advise for years. Unfortunately, 30% of companies report that they still do not have a disaster recovery program in place, and two out of three feel their disaster recovery plan has significant vulnerabilities. If you are one of this 30%, NOW is the moment to put at least short plans together, and if you are one of the two out of three, NOW is the moment to review your plans and make all the necessary changes.

For many business owners, the differences between a Business Continuity Plan and a Disaster Recovery Plan are not very clear. They are actually extremely related. A Business Continuity Plan is more comprehensive. It is about the means by which loss of business can be avoided and defines the business requirements for the Disaster Recovery Plan. A Disaster Recovery Plan, on the other hand, specifies the processes through which you are going to resume business after a traumatic event, for instance a direct hit from a hurricane.

Ideally, you should have one of each not only written down but also communicated to all stakeholders, and fully understood and PRACTICED by all personnel. Since life is very seldom ideal, let's just say that you should have A plan, whatever you want to call it, telling you and all your stake holders (employees, clients, vendors, etc...) how you are going to deal with disaster when it strikes.

It is also important to have a plan that your personnel will understand and comprehend. Your people need to be able to express their ideas and concerns. Let them know the entire plan, what to expect, how the business will continue to grow and how that is going to transpire. Go through each step of the plan with them. Do "what ifs" and "in case of" so that everyone is on the same page. Train your people so that the plan becomes second nature. Train; train, again; and train a third time. Test through practice drills; test, again; and test a third time. Practice; Practice, again; and Practice a third time. Remember that you and your team will have to execute the plan under stress so the more automatic it has become for all, the higher the chance of success. And when in doubt, train, practice and test!

The first step in your plan preparation will be to identify the activities most critical to meeting your business objectives and list them. Your second step will be to identify the needs of each activity to support business continuity, or at least recovery. Of course, your exact needs will depend on your specific business, but here are a number of requirements shared by most businesses: facility, employees, paperwork and corporate data.

Alternative Location Plan

Have an alternative location plan written and implemented. If your present facility receives structural damage in a hurricane, you and your employees need to know what to expect after the storm. Your plan should therefore include more than just the alternative location. Everyone needs to know and understand what is to be expected. You and your employees need to discuss where everyone will park, the phone system, how many computers will be used, who is responsible for what, etc. No detail is too small. Include everyone this plan will affect. This is not something that should be put off. There is a lot of planning that needs to be tweaked, worked out and

understood. The more you are prepared, the less stress you will have if and when the threat of a storm is upon you.

Employees

You need to prepare a communication plan for immediately after the storm so that everyone can rapidly and easily inform you of how they have fared the storm. A great way for you to be able to contact your employees after a hurricane is with a toll free 800 number. These numbers allow you to leave messages, retrieve messages and conference call. This allows you to make sure everyone and their families are safe. When I was doing the research for this book, I came across a few websites that offer what seems to be a great solution for companies that are interested in this type of service. Please understand, I am not recommending these particular companies; I am just making you aware of these types of services and giving you and your business a starting point: www.nocostconference.com, www.easycall.net and www.freeconference.com.

Make a Plan detailing who will do what if a hurricane threat happens. The old adage, "Two heads are better than one," works really well in this instance. When making your "In House" plan it is best to give duties to each person. Be sure everyone is aware of not only their own duties but also the obligations of every other person. This way if for some reason a person is out of the office, another person can step in to complete the tasks assigned to the person who is out.

As with the rest of the plan, this part should be written and a copy distributed to all employees and, eventually, other stakeholders, if needed. Make sure each business critical task also names a fail-over person. You wouldn't want your whole plan to fail because one critical person cannot get to your facility as soon as needed.

Paperwork and Important Data

This part of the preparation is cumbersome, but it is absolutely necessary. You need to keep an up to date copy of all your critical paperwork in a safe location. This includes all contracts (including clients, sub-contractors, employees, lease and vendors contracts), contacts (up-to-date clients, vendors, and employees), accounting (tax information, payroll spreadsheets, credit card information, detailed Accounts Receivables and Accounts Payable information), utilities (water, electrical, internet provider and telephone), deeds and insurance information. This information is imperative to keep your business running. It should always be kept up to date. You also need to have a backup of all this information off site in a place where you can retrieve it at any time.

Make an up-to-date list of contact information of all your clients' address, phone number, websites, email, alternative phone numbers, faxes and any other information such as accounting services, supplies they need to keep their business running, previous work done, contract information and future needs. Talk to them as to what to expect should a hurricane hit your business. Put them at ease with your plan of action and how it will be implemented with their needs in mind. Be sure they have the contact information and alternative contact information for you and your company.

Do not forget to keep an up-to-date list of contact information of all your existing leads and prospects. The client information list will allow your business to survive, and the leads and prospects information list will allow your business to prosper.

Make copies of vital office records and store a copy outside your office premises. Accounting databases (including tax information), insurance records, and payroll spreadsheets are just a few of the records you should make copies of and store in a safe place. You will want to consider where you will store these records. You do not want them stored in a facility that has the propensity to also be affected by a hurricane.

Make an up-to-date list of full contact information for all key personnel. Address, phone number, alternative phone number, social security, wage information and emergency contact phone number. Be sure these same people also have contact information for you, too. Make an up-to-date list of all suppliers/vendors and their contact information: Addresses, contact person, phone number, website, email, fax number, items you purchase from them and any notes that are necessary.

Compile a list of alternative suppliers. If you use any local suppliers, they will probably be affected by the same storms that will be hit you. Get acquainted with the alternative vendors, and fill out any necessary paperwork with them to save time should you find it necessary to use them.

If your company does not own a firebox, you should purchase one now. When choosing which firebox to purchase, keep in mind what it needs to hold. For example, make sure it is big enough to hold a three ring binder containing the data critical to the functionality of your business: deeds, titles, insurance contract, client contracts. Designate one person, other than yourself to share the responsibility for grabbing the firebox. Whoever this person is, it is important they also have the key to open the box should you need the information from the box.

When making the binder with key information for the firebox, you should also make additional copies of important data to be placed in binders and given to key personnel. In these binders keep the information relevant to the people and their positions. Give them the responsibility of keeping this data safe. Just remember, if the data is important there should be a master copy of it in the binder that goes into the firebox.

Business Assets Inventory

Have a complete inventory of all your business assets. This includes your computers, network devices, phones, desks, file cabinets, chairs, refrigerator, degrees, supplies, storage bins, lunchroom items, restroom items, racking, bookcases, conference room furnishings, decorative items, waiting room items, fax machines, personal items, software, copiers, customer's items, lamps, event items, books etc.

Everything in your office and/or store that is either valuable or fundamental to keeping your business running properly should be cataloged. Be sure you have pictures, model numbers, serial numbers, make, when each item was purchased, how much you paid for it, whether it is still under warranty, a copy of the receipt, color, description, and any notes you deem necessary to explain each item. It is very important to have this very detailed and complete inventory because if your roof caved in during a storm and everything you had was lost, during this stressful time it would be impossible to remember everything you had. Once you have this completed inventory remember to add any new purchases to the list. The same rule applies if you replace any item, be sure to remove it from your inventory.

Ideally, you should get a professional inventory made by a specialized third party, like Florida Inventory Services (http://floridainventoryservices.com). Specialized third parties work in conjunction with insurance companies and know exactly what information you need about each item to ensure the best possible results in case you have to claim a loss. Third party inventories also have a much higher probatory value than the inventory you make yourself because of their neutrality.

Insurance Coverage

Evaluate your insurance coverage. You need to know if you have ACV, actual cash value, or replacement cost insurance. You may find after you completed your assets inventory that you need additional insurance to cover everything. Be sure you understand your insurance coverage especially when it comes to what is covered and not covered when it comes to a hurricane. There could be additional deductible money for losses during a hurricane. Make sure you talk to your insurance broker or agent during this preparation time and you fully understand the extent and conditions of your coverage.

Before the Storm Business Specific Brainstorming and Note Summary

Section 3: Technology Specific Information

Network Documentation

Another item you should already have but most likely do not have is full documentation of your network. This should include Internet Provider information, a plan of your connections, full information about the structure of your internal network, public and private network and devices addresses, server(s) specifications and configuration, router and modem specification and configuration, printer specification and configuration, password policies, etc…

This is information you need to have at hand at all times so that any I.T. person asked to work on your network can know at a glance how things are setup instead of having to spend time on your clock just understanding and/or remembering how the network functions. In times of possible disaster this information is even more necessary and could make the difference between a quick and painless network recovery and a long and costly network down time. Ask your I.T. provider to put together the information, give you a hard and a digital copy and keep an additional copy in their records. Remember to put a hard copy in your firebox with the rest of your business critical information.

If you cannot or do not want to have a full, formal network documentation binder, at the very least, make sure you gather ALL user names and passwords. Make a list of them and put a copy in the firebox. Also make sure you have an accessible copy should you need a way into your data from your computer or computer network. This is something you should already have in place. If you do not, now is the time to gather all of your information. Once you have it, it is a very good idea to ALWAYS keep it up to date.

Be sure all of your network devices are up to date in your asset inventory. Be sure to have description; model number; serial number; photograph; copy of the receipt; warranty, if any; replacement cost and any notes that are important if something should happen to your devices.

Safe Location For Electronic Devices

Within your office, you need to decide the safest place to put your computer and other technological devices to keep them out of harms way. You must keep in mind what the winds, rains and tornados of a hurricane can do to a building. It is more than just a window getting broken. Water can come up through the foundation, and wind can uproot a tree and drop it on the roof of your office. For many helpful hints be sure to watch the bonus video about choosing a safe location for your computer and other devices on the book's website http://www.protectyourbusinessfromhurricanes.com.

Look for the space that is least likely to receive damage. A closet without an outside door or window may be the right space for you. This is true even if you board up your windows and sandbag your perimeter. Sometimes, a closet or a restroom is the safest and most secure place to put your electronics. Sometimes, your office or shop does not have the perfect place to store your precious electronics, just do the best you can with what you have.

Before the storm is also the best time to talk to your I.T. service provider. Make a plan of action as to what will happen if the threat of a hurricane does occur. Who will be doing the tearing down of the computers and network devices to keep them safe from the storm? Is it you or your I.T. service provider that is taking on that responsibility? Make your plan detailed and understandable to you, your staff and your I.T. service provider. Everyone must understand their role and who is accountable for what. Talk to your I.T. service provider about your safe space and what you are using to keep your equipment off the ground. Discuss what will happen after the storm. Will they be the ones setting the equipment back up? Do they know about your alternative location and where it is located? Make your plans now and make sure everyone who is involved understands and comprehends how the arrangements will be carried out.

Bagging Computers and Electronic Devices

If a storm comes your way and you have to leave your facility, you will have to bag your computers and all technological devices to protect them as much as possible from the elements. The first thing you need to do is to designate the employees who will be in charge of the bagging. It is best to have them work in teams of two people to avoid taking risks with the equipment. The number of teams will depend on the number of safe locations you can use and the number of computers they will have to take care of. Generally speaking, count one team and ten computers per safe location.

Have your teams practice the bagging several times until they are certain they can get it done safely and right in a timely manner. For many helpful hints be sure to watch the bonus video "How to wrap up your electronic devices, and protect them as much as possible from hurricanes winds and rain" on the book's website:
http://www.protectyourbusinessfromhurricanes.com.

You may be thinking why would I do that when everything is fine and working properly. It is best to do a couple of dry practice runs so when the time comes that you do have to get ready for a hurricane, your teams will know what to do, and there will be less stress. Have them keep the cables for each computer or network device together and wrap them neatly to keep them from crimping or getting any damage. The dry practice runs do not have to involve all your computers and devices but it should be repeated enough times that the employees in charge feel completely at ease with the whole procedure.

Have your teams take pictures of all equipment and label it. Make sure they use names that make it easy to recognize which computer is which. Try not to use employees' names but rather locations, (first office right corridor) or functions (accounting) names. Have your teams practice unhooking your computers and network devices. In your chosen "safe space" have them prepare a table, desk or anything strong enough to store your devices. You will need to store your electronics off the floor in case of flooding. Be sure the table, desk or racking can hold the weight of all these devices.

Have your team members wrap up each device as shown in the video and place it on the table or desk. Then, have them protect the devices against rains and pieces of ceiling plaster by placing another table, a piece of plywood, or simply cardboard on top of them.

If you do not feel at ease with having your people do the unhooking and bagging in house, you will need to get in touch with your I.T. service provider and let them know that, should a storm come your way, you will be needing their help to protect your equipment.

Backups

Whether you own a SOHO (Small Office Home Office), a small, a medium or a large business is immaterial. Your data is your most important asset and should be protected at all times as if your life depended on it… because it does. At least, the life of your business, which represents your and your employees' livelihood, does depend on the survival of your data.

National statistics show that 93% of companies that lost their data for 10 days or more filed for bankruptcy within one year of the disaster, and 50% filed for bankruptcy immediately. (Source: National Archives & Records Administration in Washington). You do not want your business to become a part of this statistic. You should already have a local and remote backup strategy in place, but if you don't NOW is the moment to set one up.

Local Data Backups

Ideally, when you backup to a local device, you will backup your whole system, including the Operating System and all the applications. It is the only way you can recover fast in case of disaster. However, if this is not possible, backing up the actual data will at least allow you to keep functioning, albeit after sometime.

Local backup devices include CDs, DVDs, jump drives, tapes and external hard drives. CDs, DVDs and jump drives are more appropriate to backup exclusively the data. Tapes and external hard drives allow you to backup the whole computer/server.

Unless you have an extremely small amount of data you want to stay away from CDs and DVDs. Recording a disk a day is extremely time consuming, so in spite of all your efforts you WILL end up missing days and Murphy's Law tells you that disasters will happen the day after a day you missed your backup. Jump drives offer more space and are more easily rewritable. This can be handy since it allows you to reuse the drive. But beware, it can also lead to a catastrophe if someone accidentally uses the drive for other purposes… and erases your backups. Please, do not be complacent and think that this kind of thing will not happen to you, because it will.

Tapes and hard drives are the most recommended backup devices because they offer enough space to backup the entire system. You want to keep at least two full copies of the system and two differential (everything that changed since the last full backup was made) copies. This will save you from complete disaster if one of your backups is not done properly or a backup file becomes corrupt. This, too, is a thing that will happen to you, believe it or not.

Back in the days before USB ports and external hard drives (Hmmm... am I dating myself?), tapes were kings. Businesses used to setup complicated tape rotation systems and backup strategies. Nowadays, many businesses, especially the small and medium ones prefer to use external hard drives. If you are going to use external drives, avoid the ones that can be bought pre-built and have your I.T. service provider build one for you. It will not be as cheap, but pre-built hard drives come with a one year warranty whereas the drives your I.T. service provider will use offer between three and five years depending on the brand and model. Since the drives come from the same manufacturers, the $1 million question is: do you really think manufacturers will be as strict on quality control on the one-year-warranty pre-built drives as on the five-year-warranty drives? Whatever your preference, make sure you choose drives that at least double (triple would be better) the size of your internal hard drives. You will need at least two drives so you can rotate them, and always have one copy of them off the premises. But drives are like friends at a party: the more the merrier. Do not forget to label your drives so that you always know which one has to be used next.

You will also need backup software. Microsoft backup software has the enormous advantage of being free and preinstalled on your system. It is, however, neither the most reliable nor the easiest to use for recovery. If you can, purchase a third party imaging software. Imaging software makes recovery easier and practically automatic. This is exactly what the doctor recommends in cases of high stress due to data loss following a man made or natural disaster.

Once you have chosen hardware and software, create a backup strategy and implement it IMMEDIATELY. Hook up the hardware and setup the software to start backing up your system TODAY and keep backing up EVERY DAY. Make sure you activate the "Verify data after backup" feature to increase your chances of avoiding bad backups.

Remote Data Backups

The second part of your backup strategy concerns remote data backups. The general idea is that you send a copy of your system, or at least of all your data, to a remote server that you access through your Internet connection.

The biggest danger businesses have with remote backup services is lack of knowledge in what to look for. There are thousands of companies offering this service, some of them for very little money. But, not all data backups companies are created equal. You absolutely need to make sure you choose a good, reliable vendor, or you'll get hit with hidden fees, unexpected "gotchas," or with the horrible discovery that your data wasn't actually backed up properly, leaving you high and dry when you need your data the most.

If your remote backup provider doesn't meet all 7 of these points, then you'd be crazy to trust them to store your data:

1. **Military-level security both at the data transfer level, and at the data storage level.** You need to make sure the company housing your data is really secure. This is no joke: we are talking about your financial information, client data, and other critical information about your company. You cannot possibly trust your data to anyone who doesn't offer the following security measures:
    a. **Make sure your remote backup provider is in compliance with HIPAA, Sarbanes-Oxley, Gram-Leach-Billey, SEC NASD, and the soon to come FTC "Red Flag Rule."** These government regulations dictate how all organizations that issue invoices need to handle,

store, and transfer their data. If you are in the medical or financial fields, you are required by law to work only with vendors who meet these requirements. But even if you think you are NOT required to comply with these regulations, you still want to choose a provider who is compliant because it's a good sign that they have high-level security measures in place.

b. **Make sure the physical location where the data is stored is secure.** Ask your remote backup provider what kind of ID system, video surveillance, and/or of card key system they use to allow only authorized personnel to enter the site.

c. **Make sure the data transfer is encrypted with the highest standards in encryption** to prevent a hacker from accessing the data while it's being transferred.

2. **Multiple geographically dispersed data centers.** Whenever we speak of data security, redundancy is the word you want to hear. You want to make sure multiple copies of your data get stored in more than one location. If one of your provider's vaults is destroyed by a hurricane, a tornado, an earthquake or some other disaster, there will still be a copy of your backup in a different city where the disaster did not strike.

3. **Make sure your remote data backup provider will eventually send you copies of your data on a DVD or an external hard drive.** If you get hit by a storm, you might not have really fast broadband Internet for a while. In those conditions, downloading your data through your Internet connection could take days, if not weeks and seriously endanger the survival of your business.

4. **Ask your remote data backup provider if you can send them your initial backup on a hard drive or a DVD.** Depending on the amount of data you need to backup, transferring it through the Internet could take days or weeks. Copying your data to a DVD or external hard drive

and sending them the drive will make things faster and more convenient.

5. **Ask if your data can be restored to a different computer than the one it was backed up from.** You don't want to be left without a backup just because the original computer disappeared for some reason, or you don't currently have access to it.

6. **Demand regular status reports of your backup.** All backup services should at least send you an e-mail if the backup failed and give you the reasons of the failure in the e-mail. The best providers also e-mail you regularly about the status of your backup, the amount of space you are using and the amount you have left.

7. **"Self-serve" is for gas stations, not for remote data backups.** Most cheap online backup services will let you download the software from their site and set it up yourself. Unfortunately, you will not be sure whether you set your system to back up correctly until it's too late. Do not put your valuable corporate data at risk to save a few dollars. At the very least, ask your service provider to have a technician check your settings to make sure you did the setup properly.

Test Restore

Any number of things can cause your backup to become corrupt. By testing it regularly, you'll sleep a lot easier at night knowing you have a good, solid copy of your data available in the event of an unforeseen disaster or emergency. You do not want to wait until your data has been wiped out to test your backup; yet that is exactly what most people do – and they pay for it dearly.

Whether you ask your I.T. service provider for their help or you do it yourself, you need to test the quality of both your local and your remote backups regularly by implementing a test restore. If your data is very sensitive and you cannot afford to lose it, then test restores should be done monthly. If your situation is a little less critical, then quarterly test restores are sufficient.

Whatever the sensitivity of your data, hurricane season is THE moment when you absolutely need to do test restores. At the very least you need to do one restore of both the local and the remote at the beginning of the season and one each time a named storm seems headed your way. You do not want to wait until you are under alert and have 200,000 things to do under high stress to check the quality of your backups.

If you need help with setting up and maintaining your local and remote backups, you can always email us at info@itbythesea.com. If you are looking for a reliable and secure remote data backup provider, check: http://datalifeguards.com.

Before the Storm Technology Specific Brainstorming and Note Summary

Part 2

IT'S A COMING

When A Hurricane Warning Has Been Sounded

Remember how we asked you to make sure you trained all of your employees then practiced and tested all parts of your plan? Well, if you have done your homework thoroughly now is the time when you will start harvesting what you planted. Planning will not allow you to completely avoid storm related stress, but it will certainly alleviate it enormously for you, your employees and your stakeholders.

Serious Silly Talk ~
  Chicken Little in a Bit of a Frenzy

There are many versions of the children's fable, Chicken Little. The story has been used for close to a century to teach the need for courage, astute observation and judgment born of fact. In the story, Chicken Little was meandering through the woods one day when an acorn fell on to her little head. Without reasoning, she rushed to a conclusion, "Oh my goodness, the sky is falling!" Then, based on her faulty conclusion, she determined a course of action: "I must go and tell the lion," she said. After all, the lion was king of the land.

In the course of her journey to tell the king, Chicken Little met her friends Henny Penny, Cocky Locky and Turkey Lurkey. Her friends were all merrily going about their business, which was that of sustaining life, when they encountered the panicked Chicken Little. When they asked Chicken Little where she was going, she told them that the sky was falling and she was going to tell the king.

At first, Chicken Little's friends met her cry, "The Sky is Falling, The Sky is Falling!" with skepticism. But, when her friends asked her how she reached her conclusion, how did she know? She answered their query with the assurance that she felt it; therefore, it must be so. With her assurance, her friends became distracted and joined Chicken Little in her journey to tell the king.

After a while, the group met Foxy Woxy. When Foxy Woxy inquired politely where they were going, they told him that the sky was falling and they were going to tell the king.

Foxy Woxy slyly told them he knew of a short cut to the king and led them to the entrance of his foxhole. It is here that the several versions of the story diverge. In some stories, the group enters the foxhole and is never heard from again. In other versions of the story the group is saved by the king's hounds when Foxy Woxy hears their barking and runs for his life.

My point in telling this story is not dependent on which version of the story is told. My point is that all of Chicken Little's friends were going about their business, that of sustaining life, when they became distracted by an imaginary danger. All of Chicken Little's friends left their tasks to follow her on her journey to tell the king as they joined in the frenzied chorus, "The Sky is Falling! The Sky is Falling!" And, because they all reacted to an imaginary danger, they put themselves in the path of a very real, life threatening danger.

CHICKEN LITTLE PUT HER FRIENDS IN HARM'S WAY BECAUSE OF HER FEARS.

A casual reading of the story reveals a likeable but not very astute Chicken Little leading her friends into harm's way because of her tendency to blow things out of proportion. She's imagining harm that is not there. She's worried about things that don't even exist. And subsequently Chicken Little and all her friends fall victim to a harm that is actually real.

In the version with a good outcome, the Lion King stops the group, inquires about the cause of the frenzy, determines reality and restores calm. This story teaches the necessity for investigation and deductive reasoning.

When a hurricane is coming in your own business story, you, as the business owner, need to become the Lion King! Don't allow a **"Frenzied Chicken Little"** to distract your staff and put you, your employees and your business in harms way.

We all know that people react differently when exigent circumstances arise. As a business owner, it is critical that you recognize this fact. When a storm is coming, you will need a functioning staff to help you prepare. One employee may be a dependable rock with muscles of steel who can shutter your windows all by himself *under normal circumstances*. But, if that employee suffers from anxiety attacks and freezes under the stress of an impending storm, you need to assign a different employee to cover the task. Surprisingly, the best person for the task might be a delicate female who becomes a Charles Atlas under the influence of an adrenaline rush.

You need to know how committed your staff will remain under pressure and stress. If a single mother is worrying about the safety of her children, it is unlikely that she will choose your bottom line over her family. I think it is fairly safe to say that people would rather lose their job than their family. Make sure you make your decision to prepare for the storm and shut down far enough in advance to allow your employees time to get home and care for their families. Don't wait until the last moment to make the decision to pack up and run in hopes of making a few extra dollars. It just isn't worth it.

On the other hand, don't pack up too early and spend all your staff's productivity packing and unpacking, packing and unpacking... There is neither a fine line nor a wide gulf between avarice and irresponsible behavior or practicing caution to a fault. The message here is if you are prepared, you need not be too fearful or too cautious.

After reading about the adorable, but not very astute, Chicken Little, is there something that comes to your mind that you may have forgotten in your own preparation?

Looking back at our four restaurant owners, they all need employees to prepare and sell hamburgers. Make sure you add employees to the list for winning the contest. Go ahead and write employees in the margin before you continue.

Are there other things that you need to add to your list? Write them in the margin notes now.

_____
_____
_____
_____
_____
_____
_____

Write Your Own Story

Now it's time to Write Your Own Story. This group activity will help you understand your employees' needs, fears, anxieties, personalities and capabilities. Listen carefully as you go through the activity. Try to discover if any of your employees will cause a frenzy that creates a breach in your plan when your staff needs to remain calm and collected as they spring into action.

Determine if there are any employee concerns that you should be aware of when you are preparing under a hurricane alert. In order to do this effectively, you need to go undercover. This can be accomplished by playing a game called "Write Your Own Story." It is important that your employees understand that their written responses will remain anonymous, that you want ideas and that there are NO BAD IDEAS because the worst ideas often give seed to the best ideas.

Ask your employees to fill out an online survey without disclosing their identity. Then, have them print out their responses, cut the questions into strips and deposit their answers in to a fishbowl. Allow room on the survey for your employees to write in some of their own questions and answers. Following are a few questions that you might consider asking:

- Have you ever been in a hurricane before? What was it like and how did you feel/react?
- Which character in the story Chicken Little is most like you and why?
- How would you respond if someone came to you in a frenzy while you were preparing for an approaching hurricane?
- Is there anything that concerns you that might cause you to become frenzied when you are responding to a hurricane alert? What is it and Why?

Be creative when developing your survey! Ponder what questions you might include that will give you insight to your employees and help you craft the best plan you can possibly create. When the surveys have been answered and the results have been deposited into the fishbowl, the party is about to begin.

Gather your staff in a conference room and tell them that they are going to help you "Write Your Own Hurricane Survival Success Story." Tell them that each of them will take turns drawing a piece of paper out of the fish bowl. They will read the question and answer, and then they will create a story line around the answer telling a solution to any problems that were mentioned or giving praise and positive reinforcement for positive answers. Each employee will build on the previous story line until all the papers have been drawn out of the fishbowl.

When the activity is finished ask your staff to review your Disaster Recovery, Business Continuity and Emergency Preparedness Plans and consider any changes they believe are necessary to make the plan better. Ask them to submit their suggestions and give them a timeframe to complete the assignment.

If you make changes to the plans, make sure to let your staff know then practice, Practice, PRACTICE. Keep your plans current and PRACTICED.

Now, you need to make sure you have everything you prepared during the "before the storm period" and haven't forgotten anything. If you have forgotten a couple of things you will have time to do them now thanks to all the good work you did on the rest of your preparation. After that, relax, BUT don't forget to keep your plan up-to-date. Then, when a storm warning or hurricane alert is issued, just do what you planned, practiced and tested during all the months that preceded the storm.

Section 1: General Information

There are gusts and sustained winds hours before a hurricane hits. Sometimes a gust can knock out power well before the hurricane hits. As a consequence, you may lose power before the storm makes landfall. Be sure you give yourself plenty of time to make the proper arrangements before the weather takes a turn for the worse.

Be sure to have your cell phone, laptop, cordless drill and camera rechargeable batteries fully charged. If you have extra batteries for these devices you will also want to have them fully charged. You may be without power for a few days so now is the time to prepare. Test everything as if your livelihood depended on it… because it might.

Gather your supplies and have them handy. These items include your tarps, nails, flashlight, lots of trash bags and extra batteries.

Put up your openings protections. This is where you will be happy you did a rehearsal before the threat of the storm. Be sure to do this well before the weather changes. Boards and corrugated metal sheets are very difficult to handle in high winds. Make sure your openings protections are secure.

Bring in any plants, ashtrays, portable or hanging signs, chairs and trashcans from outside. These items can become lethal weapons in the wind of a hurricane.

Dust off your battery operated radio and place fresh batteries inside. Test.

Make sure you have your firebox with all important information in it. These papers should include your insurance policy, contact information for clients, vendors, substitute vendors, personnel, alternative location, accounting database with payroll spreadsheet, tax information, business asset inventory and any other information you deem necessary that you will need to keep your business running after the storm.

Get fuel in the containers to run the generator. If your generator runs on gasoline, you may have a difficult time finding a gas station that still has gas. Be sure to get your gas as soon as you hear of a threat of the storm or beforehand if you have a safe place to store it. You might consider storing fuel at your residential property if you are prohibited from storing it at your business location. Propane can also be hard to find if you wait too long. People use propane in their grills to cook their food when they do not have power. Do not procrastinate when getting the fuel for your generator.

Take pictures of the outside of the building. These could be helpful if your building receives any structural damage from the hurricane.

Pilots Use Checklist And So Should YOU!

Every time a pilot gets ready to fly an aircraft, the pilot does a preflight inspection to make sure that the airplane is safe and flight-worthy. Each airplane has its own checklists that stay with the airplane. When the plane is ready for a flight, the pilot gets out the checklist and goes over each item on the list. It is systematic; it is routine. Nothing is left to memory. Nothing is left to chance. Every important detail is on the checklist and every detail is checked off the list before the pilot takes to the air because if there is a malfunction, the pilot can't just pull over to the side of the road and call a tow truck.

Preparing for an approaching hurricane is not like driving a car. It is like flying an airplane, and you are the pilot. There is no room for error. The survival of your business depends on you. Don't leave your pre-storm preparation to memory. Don't take a dangerous chance. You should be using your preparation checklists to make sure that you don't overlook a critical step in your pre-storm preparation.

Section 2: Business Specific Information

Make sure the information on your checklists is current and up-to-date. Checklist should have a note telling when it was last updated. Be sure that you are using the most recent ones. Is the contact information correct for your key personnel? Does it include address, phone number, alternative phone number and any other information you need to keep your business running after a disaster?

Check one last time to make sure all employees comprehend perfectly how things will operate should a hurricane hit the business.

Make sure you have all necessary up-to-date contact information for your businesses clients. Address, phone number, websites, email, alternative phone numbers, faxes and any other information such as accounting, services, supplies they need to keep their businesses running. One copy should already be stored off site, but you should have an additional copy in your firebox. Tell your clients one last time what to expect should a hurricane hit your business and how you will take care of their needs. Make sure they have the contact information and alternative contact information for you and your company.

Don't forget your leads, they will ensure your business has a future. Make sure you have all necessary up-to-date contact information of your leads and prospects.

Make sure you have an up-to-date list of all suppliers, their contact information and your account numbers, user IDs and passwords. One copy should already be stored off site but you should have an additional copy in your firebox. You also may want to contact your vendors to remind them of the plan of action in dealing with them after the storm. Since you (should) have already discussed this plan with them well before the storm, this conversation or email should be quick and easy. Make sure you also have a list of alternative suppliers just in case some of your own suppliers get hit with a hurricane. Again, one copy should already be stored off site but you should have an additional copy in your firebox.

Check to see that you have all necessary copies of vital office records in the firebox. Make sure you have a second copy of all necessary corporate papers stored outside your office premises and preferably in a facility that does not have the propensity to also be affected by the hurricane. (Accounting databases, insurance records, payroll spreadsheets are just a few of the records of which you should have additional copies stored in a safe place.)

Confirm that everyone understands what to expect after the storm. "If this happens, we will do this." It is very important you run through these scenarios again. Phones, internet and electricity can be down right before the storm and at times can stay down a few days or even a week or two. That is why it is so critical you talk about what could happen one last time NOW!

Check to be sure that you have an up-to-date contract portfolio in hand. This should include all contracts such as clients, sub-contractors, employee, lease agreement and vendors. You should have one of these stored off site but this will be part of the binder that is in your firebox. Just make sure that it is there when you walk out the door with the firebox in hand.

You should already have your accounting database in the binder found in the firebox. Check to see the last time it was updated. If necessary, bring this information up to date one last time.

Just like with all of your business data, your payroll spreadsheets should have any changes made to them to keep them current. If they are not up-to-date, now would be the time to bring them current.

You should already have all you and your employees' user names and passwords. The information should be stored in your information binder located in your firebox. Take a look at it to see when the last time it was updated. Are you positive that all the information is accurate? Would you bet your life on it?

Check to see that your business assets inventory list is current. You should have any new items you have purchased for your business already in the inventory. Everything should be current. Also make sure that the CD or binder containing your inventory is in the firebox.

Use the checklist in the appendix to make sure you are not forgetting anything critical. If you have forgotten something, make sure you take care of it before you leave the office.

I'm sorry that this may seem redundant. It is! However, if you have done your homework, if you have put procedures in place to keep your paperwork updated, and if your employees have followed the procedures that you have put in place, this process shouldn't take very much time at all. The procedure should be like the pilot's routine preflight checklist. If you have done your job properly, no major overhaul of your records will be needed at this time.

If you haven't already established procedures to maintain your records and keep them up-to-date, do so immediately. Note: Immediately means NOW, while you are reading this book, NOT NOW when a hurricane is on its way.

Develop a habit to check on a routine basis to make sure that you and your employees do your jobs and follow established procedures to keep your records up-to-date because you will

_____

_____

_____

_____

not have time to do a major record keeping overhaul when the hurricane is on its way. This is a very serious business. Your livelihood, the life of your business depends on how well you follow through to maintain the information in your firebox.

When a Warning is Sounded Business Information Brainstorming and Note Summary

_____

_____

_____

_____

_____

_____

_____

_____

_____

_____

_____

_____

_____

_____

_____

_____

_____

_____

_____

_____

Section 3: Technology Specific Information

Make sure all your corporate data is backed up. Make sure the local backup devices have been or will be taken off site. Make sure the remote backup functioned and note the date of the latest one. Get in touch with your remote data backups provider and let them know that you are under storm alert and that you will be offline for a few days.

If you feel uncomfortable about taking apart your computer or network system, you should have already spoken with your I.T. service provider to let them know you are in need of their services. Call them as soon as possible because the I.T. businesses are likely to have many clients in your same situation and will need to plan their scheduling as soon as possible. If you can get a call into them sooner, you have a better chance to get the help you need in a timely manner.

While your I.T. service provider is on site, take the time to communicate the "what ifs" for after the hurricane. You need to be clear on what you expect the I.T. service provider to do after the storm has passed. Talk about how you will be communicating. Be sure you have all of their contact information and they have yours.

If you and your employees are going to take your computer and network system apart yourselves, get your camera ready. "A picture is worth a thousand words," is a true statement. The photo can also be very helpful if you are not sure where the cables go for your computer or other network devices. Name each of your computers. Use names that make it easy to recognize which computer is which. Try not to use employees' names but rather locations (first office right corridor) or functions (accounting) names. Please make sure all the employees in charge of labeling the computers are in sync and respect the naming policy so that all computers names are kept consistent. Write the computer name on an index card or label, attach the name to the back of the computer and take a picture of where the cables live before you start unhooking everything. Be sure to print a hard copy of each picture, write the name of the computer pictured on the back of the picture and put it in your firebox with the rest of your hurricane packet.

Now you can begin unhooking your computers and network devices. Keep your cables for each computer or network device together. Wrap them neatly to keep them from crimping or getting any damage.

In your chosen "safe space" prepare a table, desk or anything strong enough to store your devices. You will need to store your electronics off the floor in case of flooding. Be sure the chosen table, desk or racking can hold the weight of all these devices.

Take the first device and carefully place it into an appropriately sized plastic trash bag. Place the coordinating cables within the same bag. Be sure there is a notification of some sort on the equipment to make certain after the hurricane it will be placed in the same location it came from. Close the trash bag tightly and secure to ensure water cannot seep in. Situate the covered device on the table or desk in your safe space. Take the next item and continue the process.

Once you have all of your electronics off the floor, covered and in your safe place you want to doubly protect them if at all possible. Cover the items with another table, a piece of plywood or something hard to protect them should your ceiling drop, get saturated and bits of the ceiling plaster fall on your equipment. If you do not have something hard, covering the devices with cardboard will help act as a cushion should parts of the ceiling get wet and begin to drop on your equipment.

Now it is time to take shelter and wait out the storm knowing that you have done everything in your power to survive.

When a Storm Warning Has Been Issued Technology Brainstorming and Note Summary

Part 3

AFTER THE STORM

Getting Your Business Back Up And Running

The storm is gone and your main task will now be to get your business back up and running as soon as possible. How fast you can do it will depend on how hard your region and your neighborhood were hit by Mother Nature, and that is something no individual has any power over. It will also depend a lot on how much planning and homework you did before the storm, and that is something you do have power over.

Serious Silly Talk ~
Assembling Humpty Dumpty

Humpty Dumpty is a charming children's nursery rhyme that teaches the need for caution and for writing down assembly instructions. It goes like this:

> Humpty Dumpty sat on a wall,
> Humpty Dumpty had a great fall,
> All the king's horses and all the king's men
> Couldn't put Humpty together again.

The first lesson that Humpty Dumpty teaches is one of caution. Problems can be avoided simply by applying a few rules:

Rule 1 - Don't let Humpty Dumpty fall off the wall if you can help it.

Rule 2- Carefully write down complete, concise assembly instructions as you put Humpty together for the first time. Then, take a photo of Humpty Dumpty when you're finished assembling him.

Rule 3 – When re-assembling Humpty Dumpty, look at his photo and read the assembly instructions. Don't be like the king's men who couldn't put Humpty Dumpty back together again.

66

One might wonder at the reason the king's men failed when trying to reassemble Humpty Dumpty. Perhaps it was because they didn't write down, keep, and then follow the assembly instructions.

Rule 4 – Focus, Focus, Focus! Don't let yourself get distracted. It is easy and natural to become a spectator of the damage the storm has done. Although it is necessary to have information on when utility services will be reconnected and when trees and other obstructions will be cleared away, thereby allowing your customers access to your business, it is not necessary to see all the news reports that are being broadcast 24 hour a day. You can look them up on the internet and they'll still be there five years later.

Getting your business back up and running again needs to be your personal Nehemiah Project

Nehemiah lived in Jerusalem at the time of Artaxerxes, King of Persia. Nehemiah's ancestors lost a war with Persia under the reign of King Nebuchadnezzar and had been living in captivity until the reign of Cyrus who wrote a decree that certain Israelites should go back to Jerusalem to rebuild the temple and city, both of which had been destroyed during the war.

Now, fast forward to the reign of Artaxerxes…

The Israelites were busy rebuilding the city of Jerusalem. Nehemiah was in charged of rebuilding the city's walls. This was not popular with the people who had inhabited the land while the Israelites had lived in captivity.

In order to stop the walls from being rebuilt, the locals did everything in their power to distract Nehemiah and to disrupt the work. The attacks on the workers and attempts to distract Nehemiah were so great that it became necessary for

Nehemiah to arm his workers with swords and spears and have half of them guard the other half while they rebuilt the wall. Focus was critical. Nehemiah's men worked from early morning until the stars appeared in the night sky and the long massive walls of Jerusalem were rebuilt in fifty-two days.

Nehemiah's focus was constant and became so renowned that two thousand years later building projects as far away as New York City have been dubbed, "The Nehemiah Project."

After the Storm It Is Easy To Become Distracted. You Need To Remain Focused!

Focus functions like putting on your oxygen mask during an in-flight emergency before you try to rescue other passengers on the airplane.

Without focus, you have no business, which means you have no employees, which means you have no paychecks and you have no business neighbors… Anyway, you get the point. Focus means you work your plan. Focus means that you put first things first, you complete the first priority tasks and then you focus on the next priority tasks until everything is in order. This does not mean that you can't have multiple employees each with a different first priority task assigned to them. In fact, if you have several employees, you should assign each of them a list of prioritized tasks.

After Humpty Dumpty has been put back together again, you will have the time and the ability to take care of lesser priorities.

Question:

What do you think your first priority task should be after you have checked to see if anyone was injured and is in need of lifesaving emergency services? _____

_____

68

Why?_____

_____

Did you guess that your first task would be to determine if your business suffered structural damage? If you did, you are right. Before you can start implementing your recovery plan, you have to know the extent of the problems you are going to be dealing with.

Your first task will be to determine whether or not your business suffered structural damage, and you will need to know the answer to this question before you talk to your employees. Depending on the answer to this question, you will take one path or another to recover from the disaster. Ideally, you will be able to rely on the Disaster Recovery, and Business Continuity plans you developed before the storm and recover without experiencing too much stress.

Section 1: General Information

If Your Business Has NO Structural Damage

Wear proper footwear and clothing when you check out any damage to your facility. Boots, gloves, long pants (if you can handle the heat), sunscreen and a hat are some of the protective gear suggested. And, for those "oopsie" moments, a first aid kit is helpful to have around.

Be very careful as there may be live down wires. Even without electricity, these wires can be very dangerous. If you do see live and or down wires, and you have phone usage, contact your local electrical company to inform them.

Clean up debris such as tree limbs, broken glass, roof shingles or tiles. Bag them if possible or use trashcans. Put the debris on the side of the road as you would your regular rubbish. Be very careful. After a storm there are roofing nails, sharp aluminum sheets, broken glass, pieces of lawn furniture and many other items that can be hazardous. Be sure to keep your gloves on when cleaning up the wreckage.

If there is no power, be sure to listen to your battery-operated radio to find out when it is safe to turn on your utilities. If you have an operating phone, you should contact your local utility company to get updated information. Be patient, after a storm the utility companies are flooded with calls regarding downed power lines, live wires and lack of power.

If you have boarded up your windows and doors, you will need to remove the shutters or boards. Leaving them up, while you have people inside the building is not recommended. Because you will probably be without power, you will want to open the windows for fresh air. It is also suggested to remove the shutters because if exigent circumstances occur, such as a fire breaks out in the building, your escape route (the windows) will be blocked. Once the windows and doors protections are removed, store the shutters where they are easily accessible. During hurricane season there may be more than one threat, and you will need to have these shutters ready.

If you do not have power, and are not opened for business for a few days, you may want to strike up the generator to run fans and other electrical items. Use your discretion here.

If Your Building HAS Received Structural Damage

Be sure to take photographs of the building and the damage received from the storm. This could be very helpful especially because your insurance company can compare these photos to the ones taken before the storm.

If you received roof damage, you may want to go up and see how bad the damage is. If you have to remove tree limbs do so. Keep in mind there may be live electrical lines so be very careful. If you can tarp the roof damage to avoid any further destruction that may occur due to additional bad weather.

It is recommended you secure your building after any structural damage. This means to try and mitigate any further damage that may be caused by additional rainfall or other weather problems. Do the best you can, just be safe while you are doing it.

You may want to keep your windows and doors coverings up. You will not be working in the building and this will help you to protect your business items from vandals and looters. It will also help deter any curious on lookers

The Federal Emergency Management Agency (FEMA) is an organization you will need to contact if you have any loss. FEMA is not only helpful to homeowners but to businesses alike. They can assist to rebuild homes and businesses. You can contact FEMA at www.FEMA.gov or by calling (800) 621-3362. For people with hearing or speech disabilities contact FEMA at (800) 462-7581.

Section 2: Business Specific Information

If your business has NOT received structural damage

As soon as you have phone, check with all employees starting with your key employees. Make sure that they are safe and that they are able to get to work. (Storm surge, tornados, and downed trees are just a few of the calamities hurricanes can cause. Different areas of the same town can receive different effects from the same storm. Some locations get hit harder than other locations.) Communicate where the business will be opening. Is it in the regular place of business or the alternative location? Find out if here is anything you can do for your employees. And remember, people handle stress differently, and they may need a little extra time to catch their breath.

Contact your Internet provider. Try and get an update as to when you will be getting your internet connectivity back up and running. Just like with all the utilities during this time, be patient when you call them. They have a lot of emergencies that are top priority. With that being said, you should be able to get some information as to what to expect.

Contact your local water company. Make sure it is safe to turn your water back on. Use common sense here. If you see a water main break down the street, you do not want to be turning your water on at the moment.

As soon as possible, contact your clients and vendors. Let them know when your business plans to re-open. If any client or vendor has been affected by the storm and is moving to an alternative location, make sure you get all the information you need about this location. Document everything and make sure all employees who need the new information receive it as soon as possible.

If any of your vendors has been affected by the storm, immediately contact alternative vendors from your list. Let them know as soon as possible what products and materials you will need and when you will need them. Remember that all your local competitors are also using alternative vendors so those vendors might have problems fulfilling all this new demand. You don't want to be the one left without critical products or materials.

If Your Business HAS Received Structural Damage

Set up your alternative location to prepare for business as soon as possible. The quicker things get back to any sort of normalcy the less stress you will encounter.

In the alternative location you may need to contact the power company, the Internet provider and the water company.

If your alternative location is a space that was for rent previous to the hurricane, these utilities will need to be turned on.

You will also want to let the utilities companies know that you are in an alternative location and approximately how long you plan to stay in it. This will allow them to suspend service in your original facility and you to avoid paying for service in two locations.

You will need to contact your insurance carrier to file your claim. You will need to have your policy number handy. Be sure to give them a contact number. You should also have your business assets inventory handy as well as before pictures and after pictures from the exterior of the building. This can help speed up your claim and get you back to business as usual. If you need help with unemployment compensation or a replacement workforce contact the One-Stop Career Center at www.floridajobs.org or call them at (866) 325-2345.

If your building has received structural damage you may want to contact Florida Small Business Development Center (SBDC) for assistance with Small Business Emergency Bridge Loan. This loan will help your business receive emergency money to "bridge" you until your insurance company or FEMA settles your claim. These loans have a quick turn around time. They are short-term loans for up to $25,000.00. Do not wait to contact SBDC if this bridge loan is something your company needs to get you back in business. There is only so much money to go around and it goes quickly. You can find your local Small Business Development Center on line at www.FloridaSBDC.com or by calling (866) 737-7232 or (850) 473-7800.

Section 3: Technology Specific Information

If Your Building Has NOT Received Any Structural Damage:

After protecting your computers and other network devices before the storm, you do not want to have them harmed by a short due to a surge when the power gets turned on.

If you had your I.T. service provider prepare your computers before the hurricane hit, you have already made arrangements with them to put your computer network back together.

Check each electronic device as you pull it out of the protected bag to be sure no water has seeped in. There are a lot of variables when it comes to the hurricane aftermath. Storm surge, a stalling hurricane, lack of drainage or a roof caved in, are just a few of the many unpredictable things that could happen. If no water has made it into the building you should not have any problems with water seepage. If in doubt, please, contact your I.T. service provider before you do anything. You do not want to waste all the hard work you and your employees did to protect your equipment by plugging in a humid device.

You will also want to make sure your surge protectors and battery backups have not received any water damage. This could be detrimental to your electronics.

Once everything is dry and successfully tested, you can start rebuilding your network. Check and double-check each connection to make sure that you only have to do the work once. Start by setting up your server(s) where it belongs, plugging it in and making sure it boots up. Then do the same with each device. Once everything seems to be working on its own, plug in your router(s) and switch(es) and power them up. Make sure you also connect them to the patch panel. Connect each device to its drop and test intra-site connectivity. To get back on the Internet, shut all router(s) down, connect your modem to either your DSL or your cable and turn it on. Once the Internet light is on steady, power your router(s) on. If nothing was hit by lightning or a surge during the storm, all computers should get Internet connectivity immediately.

Once you are back on the Internet, connect your local backup back to the server. Then get in touch with your remote data backups provider and let them know that you are back online and when you will be resuming your backups. This is important because your remote data backups provider may very well have locked the access to your account to protect your data whilst you were offline. You do not want to have your first backups refused, and suffer an eventual loss of data after you so successfully survived a storm.

If Your Office Building HAS Received Structural Damage.

Depending on how much damage your business building has received, you may want to work out of your alternative location.

If you had your I.T. service provider tear down your computer network, you will have already spoken about what will happen after the storm. You should contact the I.T. service provider and have them hook up your network system in the new location. Make sure they document the design of this "new" network and the Internet Provider information. Keep this information in the binder with your original network and Internet Service Provider information so that they will be at hand if you have a problem and when you need to cancel your accounts to move back to your own facility.

If you did the tearing down in house, and the employees in charge of doing it documented everything with labels and pictures as recommended earlier, they should be able to put your network back together on their own. They should not have problems with internal connectivity since those settings will not have changed. If your employees have trouble with Internet connectivity, remember that your Internet connection is different in the alternative location and you most likely will have to make changes in the settings of your main router. Most modern routers have user-friendly windows and are relatively easy to set up. If you do not trust yourself or your employees to change the settings, call your I.T. service provider and ask them for help. A short visit to your alternative facility should allow them to finish your setup. In fact, they might even be able to guide you over the phone.

After the Storm Brainstorming and Note Summary

CONCLUSION

Serious Silly Talk ~

All is Well… Or is It?

Most people are familiar with Charles Dickens' book, *A Christmas Carol*. In the story business owner, Ebenezer Scrooge, is introduced as a cunning, heartless businessman who lacks a clear understanding of his employees and those he interacts with in his community. He sees Christmas as just another business day and lacks sympathy for the plights of those around him, never dreaming their troubles could affect him.

Then, the night before Christmas, he is given an opportunity to change the course of his life when his deceased business partner, Jacob Marley, comes to him in his bedchamber wearing chains. Marley gives Scrooge a warning about his future if he doesn't change his ways, then tells him that he will be visited by three ghosts during the night. Afterwards, he is visited by the ghosts of Christmas past, present and future. These visitations allow Scrooge to experience a mighty change of heart. He learns lessons from the events of the past, sees the effects of past decisions on the present and realizes the impact he can make on the future.

You may not expect to have a Scrooge experience like the one described in A Christmas Carol. However, all of us can learn lessons from this story that might just save our businesses. So, let's explore our own Hurricane Carol and visit with the ghosts of hurricanes past, present and future.

The Ghost of Hurricanes Past:

First, our visit with the Ghost of Hurricanes Past will take us to several scenes from Hurricane Katrina, the worst hurricane on record in the last 100 years.

Scene 1 - A television news weather forecaster is pointing to an image of a hurricane that is so massive that the image literally fills the Gulf of Mexico. (You can find these images on the internet by Googling "Hurricane Katrina.")

The storm is approaching the gulf states of Louisiana, Mississippi, Alabama and Florida. The size and strength of the storm is unprecedented in recent history. The hurricane is definitely coming and the massive storm will hit everyone. The only unknown is which part of the coast is going to take the biggest brunt from the storm. As time progresses, it becomes evident that the eye of the storm is heading for the city of New Orleans, LA.

The greater New Orleans area is situated below sea level on the banks of Lake Ponchartrain and the Mississippi River. Flooding is always a problem when it rains there, even under normal circumstances. A series of canals with pumping stations are strategically located throughout the greater New Orleans area to help control the flooding. The pumping stations are not even adequate for normal rains and everyone knows it. There are levies protecting the city from the waters of Lake Ponchartrain. The levies are weak and everyone knows it.

There are only two interstate highways and two U.S highways exiting inland from New Orleans. I55 runs north inland on a narrow strip of land between Lake Maurepas and Lake Ponchartrain. I10 exits to the north across a bridge over Lake Ponchartrain and to the west towards Baton Rouge. U.S. Highway 90 traverses a series of bridges over canals then follows the river until it crosses a bridge between Lake Saint Catherine and Lake Ponchartrain. U.S. Highway 11 crosses Lake Ponchartrain on a bridge that is 23.87 miles long, the longest bridge in the world. There are only four major avenues for escape three of which are extremely vulnerable to disaster during a hurricane the magnitude of Katrina. Knowing the above-mentioned facts, as all New Orleans residents do, one would have to ask, "Why is there a scene two?"

Scene 2 - A T.V. news anchor is standing in front of the New Orleans Saints' Superdome. It is one of the largest structures of its kind in the world. It's a 27-story windowless building with a seating capacity of 76,000 and a computerized climate-control system that uses more than 9,000 tons of equipment that requires power. Unfortunately, it is now filled with tens of thousands of refugees and none of its equipment is working. It has become a hell on earth and the refugees have no escape. They are without plumbing, food, medical help or, in many cases, even their own medications such as insulin. People are dying.

It is easy to understand that some people lacked the resources to escape on their own, but tens of thousands??? Why didn't the city act sooner to avert the disaster? What were they thinking?

Scene 3 – T.V. news anchors in helicopters are flying over flooded areas where people are stranded on rooftops. SOS messages stating "The water is rising!" or "We need food – help!" are written in big bold letters on streets and housetops. Why were those people still there?

It seems unlikely from the evidence presented in Scene's 2 and 3 that the city of New Orleans, or other culpable parties had their Disaster Recovery, Business Continuity and Emergency Survival Plans in place or kept them current, nor had they run through their preparation check lists or had repaired their facilities before the storm. All parties involved had been around long enough to be aware of the dangers but had chosen to procrastinate the day for preparation. Imagine what could have happened had they been prepared.

Scene 4 – A grateful group consisting of hundreds of people are being welcomed into the homes of church members they have never met who live inland from the storm. They belong to a domination that has set up a program with procedures that assigns members to look after each other in the event of an emergency.

Very early in the morning before hurricane Katrina hit New Orleans, this church organization sprung into action. The head area leader made a call to the leaders of the several congregations, who then made a few calls to the leaders of the unit's organizations, who all made a few calls telling the members that they had to evacuate NOW. Then, each member called other members over whom they were assigned and asked if they needed help evacuating.

In about an hour every member was accounted for, help was given where assistance was needed. The members grabbed their 72-hour go bags (something that their church counseled them to have on hand at all times). And all members had crossed the bridge and were evacuated to safety hours before Katrina hit land. They all avoided the danger, the panic and the traffic congestion that stranded the residents who were left in the greater New Orleans area.

Later, this same group of people and their inland hosts were seen wearing "Helping Hand" tee shirts as they passed out emergency food, water, sanitation kits and clothing to the victims of the storm while others were busy using chainsaws to clear away the storm's debris.

The point is: If a church group can successfully plan and prepare in advance to handle disasters like Katrina, so can a business, neighborhood or community.

The organization: This group started by assigning members of their congregations to check in on a few other members of the congregation and report back to a leader. The local leaders

80

then reported back to an area leader who was in charge of a much larger geographical area. Because the same network was in place in alternative locations inland from the storm hundreds of refugees were immediately met with open arms and warm homes to go to.

Although there are several lesson to be learned from the ghost of hurricanes past, focus on the fact that **YOU can choose which scene your business will be a part of**. The key is advanced planning and thoughtful preparation.

Start with your employees. You need them as much as they need you. Get organized, prepare and rehearse. Be diligent in keeping your Disaster Recovery, Business Continuity and Emergency Survival Plans current. But don't stop there. Share your plans AND this information with your customers, clients and vendors. You need them to succeed, and they need you. You are interdependent!

Questions:

Based on your current plans and preparation, in which scene would you and your employees have been a participant? _____

_____

If you had to implement your plan during Katrina, would it have worked? Why or why not? _____

_____

_____

Did you have what you needed to weather the storm before it hit, or did you show up to the store when they were sold out? Why or why not? _____

_____

_____

Did the looters break into your store? Why or why not? _____

_____

_____
_____
_____
_____
_____
_____
_____
_____
_____
_____
_____
_____
_____
_____
_____
_____
_____
_____
_____
_____
_____
_____
_____
_____
_____
_____
_____
_____
_____
_____
_____

Would you have been able to recover your critical data and successfully start up your business in an alternative location? Why or why not? _____

_____

Did you allow yourself adequate time to get you and your employees out of town? Why or why not? _____

_____

If you are not satisfied with the answers to any of these questions how can you alter your plan to make it work for you? _____

_____

Who is in your network of friends, neighbors and business associates? In the margin, make a list of the people with whom you would like to share this information.

Is your network sufficiently large and geographically dispersed to rescue you in the event of a disaster the magnitude of hurricane Katrina? _____

If not, are there others that you can invite to join your network? _____

NETWORK WITH YOUR NETWORK. Are there some professional groups or neighborhood groups that you can join? _____

_____
_____
_____
_____
_____
_____
_____
_____
_____

Brainstorm with your staff to see how you can expand your network.
Your business survival may depend on it.

_____

_____

_____

_____

_____

_____

_____

_____

_____

_____

_____

_____

_____

_____

_____

_____

_____

_____

_____

_____

_____

The Ghost of Hurricanes Present:

Next, take a trip with the ghost of hurricanes present. A hurricane is coming your way. You only have a few days to prepare. Will your plan work? Will your plan work for your employees?

If you see the day before a hurricane as just another business day and you aren't keyed into your employees' needs then you might have Scrooge Syndrome. Your employees have many responsibilities. Their job supports their life. If you fall into the habit of believing they live to work, you will most likely end up with an employee shortage when the storm is near at hand. If, however, you have taken into consideration the time it will take your employees to take care of their personal and family needs when you developed your plan of action you will have (and attract) loyal, hard working, productive employees that will safeguard their jobs, your company and your bottom line.

The Ghost of Hurricanes Future:

Finally, peer into the possibilities you can create as you visit with the ghost of hurricanes future. **The future will become what you make of it**. Creating and maintaining a business takes a lot of hard work. You have spent time cultivating good clientele and solid business relationships. However, if these clients and business connections fall victim to the storms you will suffer the consequences of THEIR lack of preparedness. Every business needs customers. All businesses are dependent, to one degree or another, on the success of other businesses.

Make sure that you share this information with those upon whose business and services you are dependent.

84

As you prepare for the storm don't forget to learn lessons from the events of the past, see the effects of past decisions on the present and realize the impact you can have on the future.

I Can Sleep When the Wind Blows!

There once was a rancher who was in need of a ranch hand. Unfortunately, there was no one around who could spare their help. So, the rancher took out a help wanted advertisement in a newspaper in a far away city. There were few inquiries, but one man showed up and applied for the job.

The man was a slightly built guy and it didn't look like he would amount to anything. He was hardly bigger than a bail of hay. The rancher worried if the man could do the job and asked him for references. The man said that he didn't have any references but that he could sleep when the wind blows.

This concerned the rancher, but lacking alternatives he decided to give the man a try. A few nights later, a storm blew in. The rancher ran to awaken his new ranch hand but was unable to do so. He finally ran out to check everything by himself only to find that the barn door was shut and bolted tight against the wind and all the animals were sheltered from the storm. Every detail had been attended to. Everything was secure. It was then that the rancher understood what the ranch hand meant when he said, "I can sleep with the wind blows!"

YOU NEED TO BECOME THAT RANCH HAND.
Can you sleep when the wind blows?

By becoming a proactive participant in hurricane preparedness, like the ranch hand, you will be able to sleep when the wind blows knowing that you have done all in your power to protect your business, your employees, clients, vendors, contacts, in short your business and personal networks when the *Hurricane Is A Coming*.

In the appendices that follow, it may seem as if there is a lot of redundancy. There is! Redundancy is required because frequently business owners and employees make preparations but fail to implement procedures, or become lax in following procedures to maintain the preparation.

It is an uncommon trait to be able to sleep when the wind blows, but it can be done.

I encourage you to TAKE ACTION to PREPARE NOW. When you take action, put a plan in place to secure your business barn door, AND put a procedure in place to keep your business barn door secure. THEN, practice, practice, practice… Make sure to check on a regular basis to see that procedures are being followed to keep your barn door secure.

When the wind blows and a storm is a coming, employees who are normally rational committed individuals will have a tendency to alter in their commitment and behavior. A mother who is responsible for part of your "It's a Coming" preparation may become frenzied and unreliable when faced with the frustration of redundancy vs. getting out the door to take care of her family.

Preparation combined with procedure applied on a daily basis will develop habits that will eliminate catastrophic failure due to lack of discipline and focus when exigent circumstances arise.

Employees would rather lose a job than a family. You must take care of your employees before they can take care of your business. Advanced preparation and excellent communication will go a long way to ensure your businesses survival.

As a South West Florida I.T. provider, I am very conscious of how much damage hurricanes can cause for homes and businesses alike. The collective information in this book is a result of years of research, education and business experience.

This book was born from my desire to help my clients and other business owners to protect their businesses from the hurricanes. To that end, I have included some forms in the appendixes in the back of this book to help you take action now. And, to make taking action even easier, I have included some bonuses that you can access on the book's website: http://ProtectYourBusinessFromHurricanes.com with the user ID: Hurricane and the ISBN number of this book as password.

Bonus # 1 is a video showing you how to choose the best location to safeguard your network equipment. Take a walk with us through several office settings to discover the safest locations in each office to protect your equipment against hurricane winds and rain.

Bonus # 2 is a video showing you how to best protect your equipment from rains and floods. Watch a tech wrap the equipment and cover it to protect it against floods, rain and possible pieces of plaster of ceiling falling on to it.

Bonus # 3 is an Interactive Emergency Response Profile. Fill it out online, save it and print as many copies as you need to ensure all your employees have the necessary emergency information on hand.

Bonus # 4 is a downloadable and printable version of the checklists that you will find in the appendices. Print them up and use them to make sure you did not forget some crucial point in preparing your business for hurricanes.

Bonus # 5 is a special for South West Florida Business Owners. I.T. by the Sea is offering you a free, no charge, no obligation Disaster Recovery, Security and Backup Audit. Since this is FREE, you have no good excuse not to do it now. If we don't find any problems, you'll have peace of mind knowing that your network is secure and that you could recover quickly in the event of a disaster. But, if we DO find a few loopholes, you'll be able to fix them BEFORE you experience an unexpected catastrophe.

I also want to be very clear that there are no expectations on our part for you to do or buy anything when you take us up on our offer.

If you are not in South West Florida but would also like to take advantage of a free audit, please contact us and give us all the necessary information about your location. I am a member of a couple of professional organizations: the ASCII group and the Association of Computer Repair Business Owners. I will personally contact one of our local members and ask them if they could provide this free audit for you.

Technology advances rapidly and new innovations become available every day, therefore, this book is a work in progress. As new information, technology and products become available I will revise this book to reflect the latest advances and information. When information is updated and revised, I would be pleased to send you an email inviting you to revisit our website so you can download the additions.

To make sure you receive the update emails, simply opt-in to our email list. We will not sell or distribute your email or use it for any purpose other than stated above so know that your information is safe with us. You will not be spammed.

My main goal in writing this book was to help you become as prepared as possible so your business has the best chance of survival. To this end I wanted to share my knowledge and this information with you. It is my hope that you will take action and prepare for the storms that will come. And, after you prepare, I hope you will share this information with your vendors, clients, employees and business neighbors. The more businesses and people prepare, the better chance that everyone will not only survive, but thrive after the storm.

If I can just believe that I brought some peace to your mind for this hurricane season, I will be happy.

If you would like to get in touch with me for any reason, do not hesitate to email me at: info@itbythesea.com.

As you proceed through the appendices, be aware of the links to information, bonuses and downloadable, printable checklists that you can share. I encourage you to send an email to all of your contacts telling them about *A Hurricane It's A Coming*. Please share this information and your emergency preparations with all your network. To learn more go to http://itbythesea.com.

Christine F. Roux

South West Florida Business Owners,

Contact Us Today for Your Free

Disaster Recovery, Security

And Backup Audit

Here's how it works...

Print and fill in the form on our website and fax it back to our office.

You can also:
>   call us at (239) 344-7574
>   fax us at (239) 236-3089
>   or email us at info@itbythesea.com

As soon as we receive your request, we'll contact you to set up an appointment for your free Disaster Recovery, Security and Backup Audit.

At no charge, we'll come to your office and conduct a thorough audit to determine:
- How fast you could be back up and running in the event of a disaster
- How secure your data is ... really
- Whether or not all of your critical data is being backed up, everyday
- Whether or not you are really protected from hackers, viruses and human error
- What steps and costs would be involved to rebuild your server and recover your data if you had to.

We will also provide you with a detailed written report in plain English that outlines where you are at high risk for viruses, downtime, or other problems, and what you can do to eliminate those risks.

Appendix 1: Before the Storm
General Checklist

- Corded Telephone
- Battery Operated Radio
- Battery Operated Drill
- Extra batteries for cell phones, laptop, drill, radio
- Generator
- Approved Storage for fuel for the generator
- Tarps
- Nails/Screws
- Hammer
- Trash Bags
- First Aid Kit
- Firebox
- Gloves
- Shoes
- Fix-a-Flat
- Know where the main water shut-off valve is located
- Know where the main breakers for power are located
- Know where the gas shot-off valve is located
- Ladder
- Window/door protection
- Alternative Location

Appendix 2: Before the Storm
Business Specific Checklist

- Business Continuity Plan – written and communicated to personnel, clients, vendors, alternative location contacts and I.T. service provider
- Disaster Recovery Plan written and communicated to personnel, clients, vendors, alternative location contacts and I.T. professional
- List of alternative locations and contact information placed in binder
- Emergency Response Plan filled up, printed and communicated to all stake holders
- Plan of employees' assigned duties when a hurricane warning has been issued
- Complete inventory of all businesses assets placed in binder
- Checked insurance coverage with insurance broker/agent and eventually rectified
- Up-to-date copy of all vital office records placed in binder
- Up-to-date copy of all key personnel's information placed in binder
- Up-to-date copy of all client information placed in binder
- Up-to-date copy of all suppliers/vendors information placed in binder
- List of alternative suppliers and contact information placed in binder
- Key personnel has your contact information
- Contact an I.T. service provider regarding your network system
- Binder with all vital information ready and placed in firebox

Appendix 3: Before the Storm
Technology Specific Checklist

- Detailed Network Documentation placed in binder
- Internet provider contact information placed in binder
- Internet contract number and specification (IP addresses, settings, etc...) placed in binder
- All electronic devices properly labeled
- Patch panel and individual network drops labeled
- Complete list of all user names and passwords placed in binder
- All computer, network and general electronic devices information is in business assets inventory
- Chosen location to protect computers, network and general electronic devices in case of hurricane threat
- Chosen personnel in charge of protecting electronic devices
- Chosen location documented and communicated to chosen personnel
- Personnel has tested locations and practiced bagging
- Local data backup checked and functioning
- Local data backup restore test done successfully
- Remote data backup checked and functioning
- Remote data backup restore test done successfully
- Reviewed alternative locations, verified and tested appropriate network cabling and drops
- Verified electrical system of alternative locations will support the charge of network

Appendix 4: It's A Coming
General Checklist

- Generator tested. Working
- Spare Approved Fuel Containers for Generator
- Extra Fuel for Generator
- Window and Glass Door Protections
- First Aid Kit
- Tarps
- Hammer
- Nails/Screws
- Battery Operated Drill
- Extra Fully Charged Battery for Cordless Drill
- Cell Phone Fully Charged
- Extra Fully Charged Cell Phone Battery
- Battery Operated Radio
- Extra Batteries for Battery Operated Radio
- Large Trash Bags
- Toiletries for Alternative Location
- Flashlights
- Extra Batteries for Flashlight
- Electricity Shut Off
- Water Shut Off
- Gas Lines Shut Off
- Electronic Devices Stored Safely
- Ladder
- Safety Rope
- Work Gloves
- Work Shoes
- Bottled Water
- Photos of the exterior of the Building

Appendix 5: It's A Coming
Business Specific Checklist

- Up-to-date customers information stored off site
- Up-to-date vendors and alternative vendors information stored off site
- Up-to-date contracts (includes sub-contracts, employees and vendors) stored off site
- Insurance information
- Deed or lease contract
- Accounting database
- Payroll spreadsheets
- Business asset inventory
- Firebox with all information
- Firebox key
- Copy of all vital office records in firebox and off site
- All personnel, vendors and alternative vendors are informed of the plan for after the storm
- Emailed clients informing them as to what to expect after the storm
- Contacted your alternative location owners
- All personnel, clients, vendors have directions to get to alternative location
- Have a key for the door of the alternative location
- Some else from the company also has a key to the door of the alternative location
- Turned off power
- Turned off gas Turned off water
- Have all keys to the business
- Business door locked

Appendix 6: It's A Coming
Technology Specific Checklist

- All computer, network and electronic devices taken apart
- Computers all named by location and name taped on the back of the computer
- Back of each computer photographed (with the name of the computer taped to the back) so you have a picture of where the cables will go when you put it back together
- Cables unhooked from computers, network and electronic devices
- Verified that cables are properly labeled to ensure fast re-installation
- Cables coiled neatly so they do not get damaged by crimping
- Computers, and network and electronic devices placed in large plastic trash bags along with the corresponding cables
- Edges of trash bag tucked under the unit to avoid any chance of water seeping in
- Each device placed on top of a table or desk in designated safe area to protect from floor flooding
- Everything covered with cardboard or another table to protect from rain and roof pieces
- Final backup of critical corporate data done
- All local backups taken off the premises
- Remote backup provider alerted of the situation